AMAZING TRAINS

Written by Alan Barnes

A Division of Harcourt Brace & Company

www.steck-vaughn.com

Trains are an important way to move people and goods from place to place. Some carry people to work. Others ship food to market. Some haul steel to factories.

Today trains move with amazing speed.
They are pulled by powerful engines.
But the first trains were wagons pulled
by horses. They moved very slowly.

Then engines were built to pull trains. The first engines used steam to move trains along tracks. The trains were mostly used to move goods. They could move as fast as horses.

The first steam trains to carry people were in England. On these trains, people sat in open cars. Each car looked like a stagecoach. These trains were the fastest way for people to travel.

Steam trains moved things quickly. These trains were called locomotives. Locomotion means "to move from place to place." Steam trains could pull a train with many cars.

Soon there were steam trains in the United States, too. A steam train named "The Tom Thumb" raced a horse and lost. After that, trains were built to move faster.

Many tracks were built across the United States. One railroad company built tracks from east to west. Another built tracks from west to east. When the two sets of tracks met, the companies celebrated together.

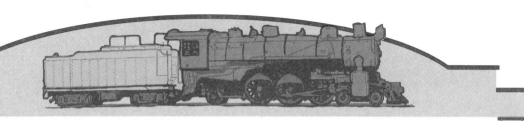

The companies pounded a gold spike into the rails to join the tracks together. Then trains could travel from coast to coast. A train trip from New York to California would take only one week. By horse, it would take six months!

People began to take more train trips. They met at the town's train station. A conductor helped them into the cars. People sat closely together in rows. Most cars had roofs and open windows.

It took a lot of work to keep the trains moving from town to town. Some workers built the railroad tracks. Others kept the trains running well. The engineers kept the trains going full speed ahead.

Soon better cars were built for trains that carried people. George Pullman helped build sleeping cars. They gave people a place to rest. Then people could take longer train trips.

Then engines were built to pull more cars and carry more people. Engines were also made to move even faster than before. Soon trains could move mail, people, or food in a short time.

Today there are very, very fast trains. They are called bullet trains. They speed along at 180 miles per hour. That is about as fast as a moving bullet.

Both Japan and France have bullet trains. Many people take them instead of driving. For the longest trips, some people ride bullet trains instead of flying.

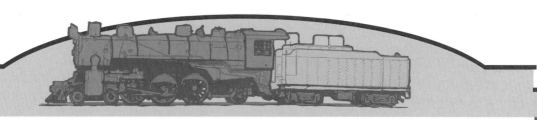

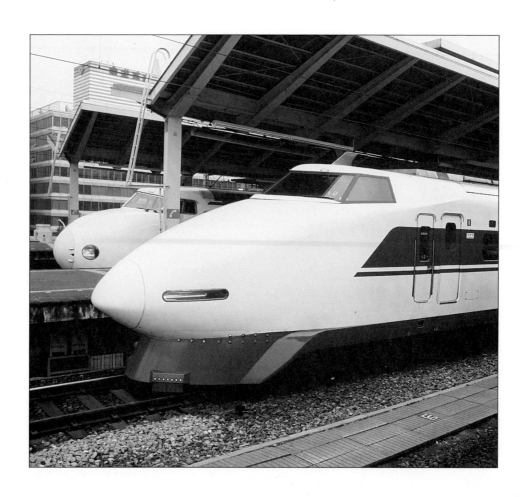

Trains are changing all the time. They are being built to move faster than ever. They are being made to carry more and to give a smoother ride. What do you think trains will be like in the future?